Fun Time Activity Book!

TABLE OF CONTENTS

WHAT ARE FUNGI?..2
MUSHROOM LIFE CYCLE..3
MUSHROOM ANATOMY 101..4
TYPES OF MUSHROOMS..5
MAKE SPORE PRINTS..6
MAKE STEM BUTT SPAWN..7
DON'T FORAGE IN THE DARK!....................................8
MYCELIUM MAZE...9
ANCIENT MUSH-TORY...10
WHAT'S IN A NAME?...12
MUSHY NAME MATCH..14
FANTASTIC WORD SEARCH.......................................15
POLYPORE PAPER...16
MUSHROOM INK..17
AMAZING INTELLIGENCE..18
PERFECTI CROSSWORD...19
ANSWERS..20
PAUL STAMETS: MYCOLOGIST & INVENTOR....21

At Fungi Perfecti®, we're dedicated to helping bees, trees, people, and planet.

We pioneer strategies to save bees using mushroom mycelium. We protect and preserve rare mushroom species—sustainably harvested from old-growth forests in the Pacific Northwest. We build soil with beneficial fungi to support plant and animal food webs—helping agriculture, forests, and ecosystems. We value scientific exploration and research on how mushrooms can support healthy systems.

It's so Mycological!

FUNGI PERFECTI® *Makers of Host Defense Mushrooms*™ | FUNGI.COM 1

WHAT ARE FUNGI?

"It's not like a vegetable and it's not like an animal, but it's somewhere in between." —*author Eugenia Bone, Fantastic Fungi*

Animals and fungi share a common ancestry! Animals are more closely related to fungi than to any other kingdom. More than 600 million years ago animals branched off and evolved systems to capture nutrients by digesting food in stomachs.

Taking a different evolutionary path, fungi developed a vast underground network of interweaving root-like branches called mycelium to gather nutrients. Fungal mycelium decomposes and recycles plant debris, filters microbes and sediments from runoff, and restores the soil.

Fungal networks are intelligent—connecting the roots of plants throughout the forest, enabling them to share nutrients and even communicate! Mycologist Paul Stamets says, "If you were a tiny organism in a forest's soil, you would be enmeshed in a carnival of activity, with mycelium moving through the subterranean landscapes like cellular waves, through dancing bacteria and swimming protozoa, with nematodes racing like whales through a microcosmic sea of life."

There are many wondrous microorganisms that exist in the earth amongst amazing mycelium! But there are a few creatures here that don't belong. CAN YOU FIND ALL 5?

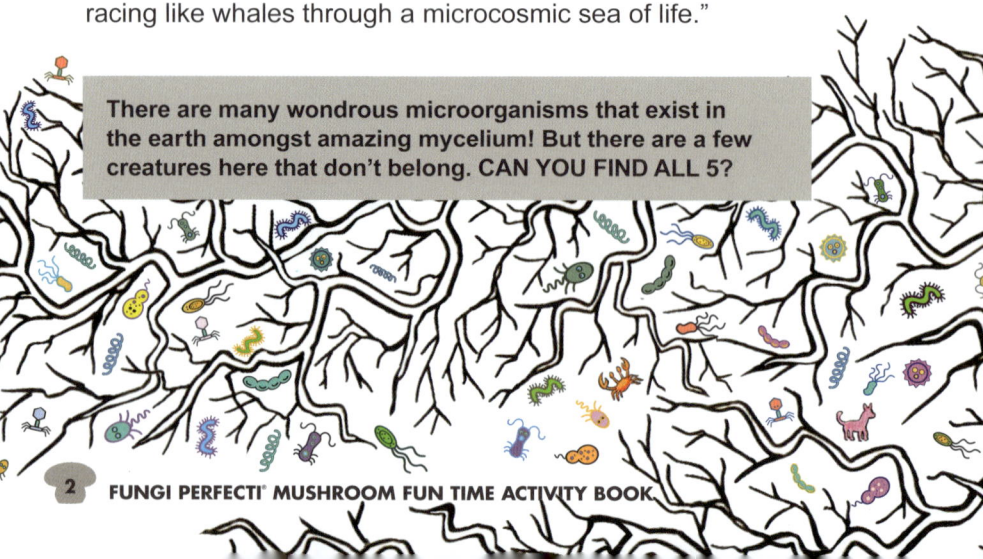

2 FUNGI PERFECTI® MUSHROOM FUN TIME ACTIVITY BOOK

MUSHROOM LIFE CYCLE

In order to reproduce, many fungi form fruitbodies, or what we know as mushrooms. While plants grow from seeds, mushrooms grow from spores—tiny dust-like specks released from a mushroom's gills or pores. When spores land on the right growing material (or substrate) it's called inoculation.

The spores germinate and start making threads of mycelium called hyphae. These mate with other compatible hyphae to form a dikaryon. Dikaryotic mycelium contains genes from both parents, giving it the ability to form mushrooms.

As the mycelium grows and branches out it seeks nutrients and creates enzymes and compounds to protect itself. The substrate is colonized by mycelium. Once the mycelium has absorbed enough food, clumps of mycelium condense into a hyphal knot. From this a pinhead or primordia forms (baby mushroom).

The mushroom matures into a full fruitbody, releasing spores to start the cycle all over again.

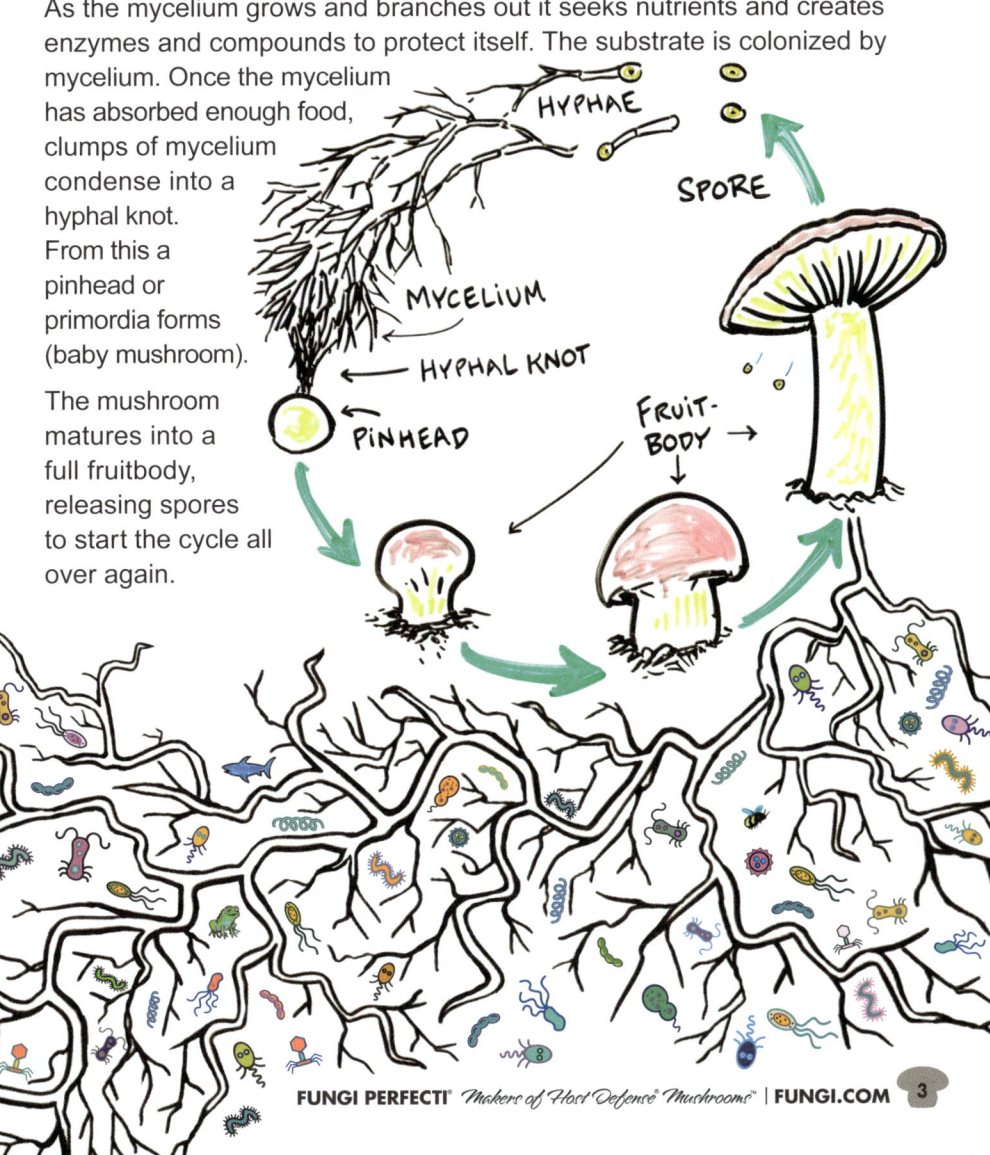

MUSHROOM ANATOMY 101

Fungi grow all over the planet and in all sorts of habitats. They can grow in the extreme heat of deserts, in soils full of radiation, and underwater in lakes and streams. Spores have survived deep in the ocean and even in outer space!

Mushrooms come in many shapes: hoof- or cup-shaped, ridged, toothed, coral-like, leafy, blobular, or little shelves sticking out from the trunks of trees (polypores). Let's look at some of the main parts of a typical umbrella-shaped mushroom:

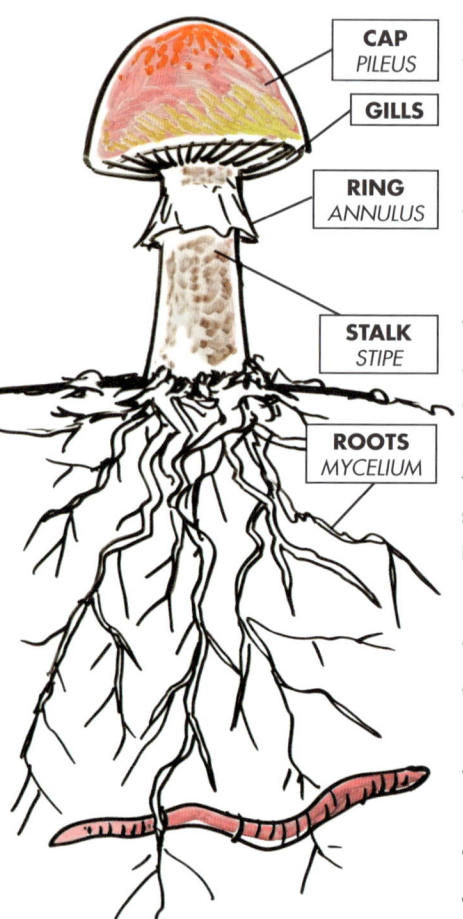

CAP / *PILEUS* / **GILLS** — The top part of a mushroom fruitbody is the cap. Gills or pores usually line the underside of the cap. That is where the spores form.

RING / *ANNULUS* — Some mushrooms have ring around the stalk that can help mycologists with identification. Sometimes the annulus looks like a ruffly skirt!

STALK / *STIPE* — The stalk raises the cap up so the spores are released into the air and can travel away from the mushroom.

ROOTS / *MYCELIUM* — Mycelium is amazing! Twining endlessly through the soil beneath your feet, a single cubic inch of topsoil can contain more than 8 miles of mycelium!

It plays a big part in the life of the forest:

- Decomposing and recycling plant debris
- Filtering microbes and sediments from runoff
- Protecting against parasites, bacteria, and other invaders
- Restoring soil
- Expanding the root systems of trees!

99% of a mushroom's life occurs during the mycelial stage. Fruitbodies are often just a small part of the life cycle, sometimes lasting for only a few days to release spores, while some mycelia can live for thousands of years.

TYPES OF MUSHROOMS

Mushrooms are categorized by how they get nutrients. Some fall into more than one category:

SAPROPHYTIC / SAPROTROPIC: THE DECAYERS

The best recyclers on the planet! Saprophytes gobble up debris like wood chips, logs, dead plants or animals, reducing them to simple nutrients for plants to use. (A roommate that cooks and cleans!)

PARASITIC: BLIGHTS OR REBUILDERS?

Though they attack and digest trees, parasitic fungi create nurse logs and increase the soil-depth—playing an important role in maintaining forest health.

Other parasitic fungi attack insects. The Zombie-Ant Fungus (*Ophiocordyceps unilateralis*) makes ants climb to a specific height in the tropical forest so the fungus can drop its spores. Don't worry, it doesn't harm humans!

MYCORRHIZAL: PLANT PARTNERS

These fungi grow mycelium around or into the roots of plants. Mycorrhizal fungi guard against pathogens and parasites, encourage root growth, and allow plants to communicate.

They can also help their plant friends adapt to drought and heat. Mycorrhizal fungi play an important role in improving soil structure and storing carbon in the form of glomalin which remains in the soil long after they die.

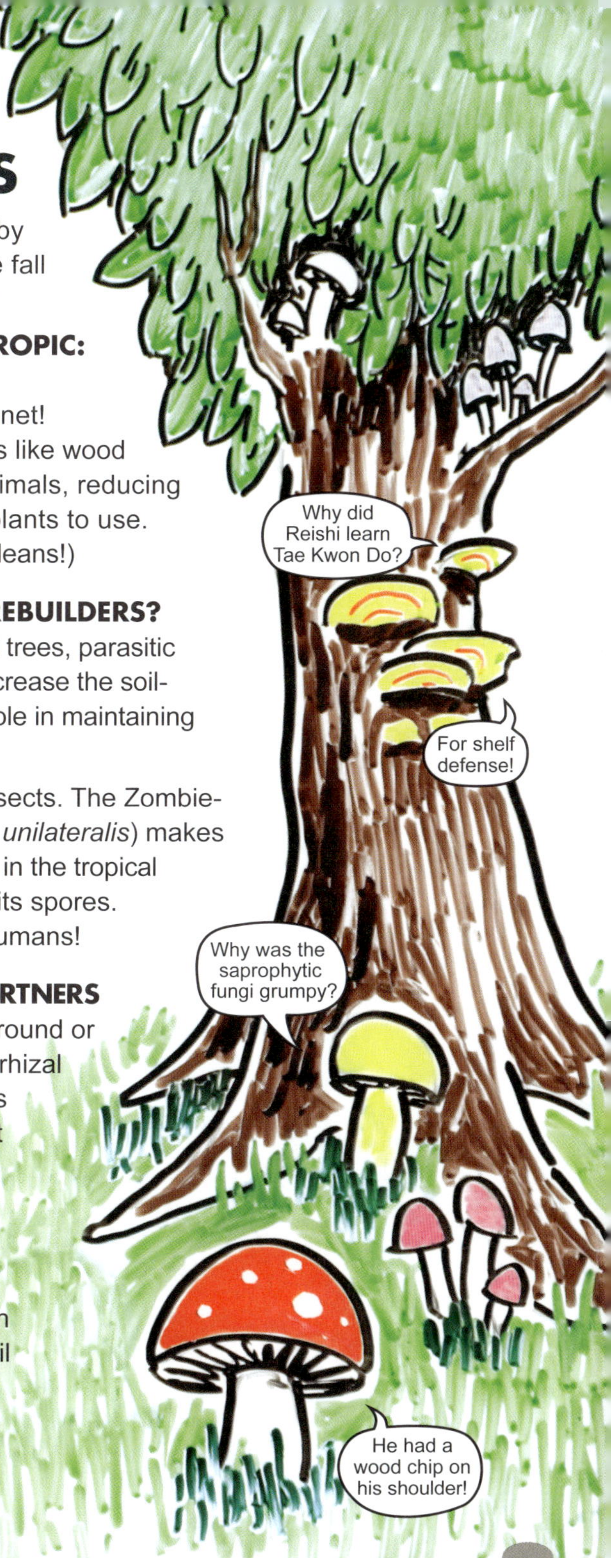

Why did Reishi learn Tae Kwon Do?

For shelf defense!

Why was the saprophytic fungi grumpy?

He had a wood chip on his shoulder!

MAKE SPORE PRINTS

A mushroom can launch billions of microscopic spores into the world, sometimes jettisoning them very far away to germinate. If you place a mushroom cap onto a piece of paper, the spores will fall from its gills and create a spore print.

Mushroom hunters use this method to collect spores for identification as well as cultivation.

To make your own spore print you need fresh mushrooms. Portobellos and other dark gilled mushrooms have brown spores which work well on white paper or paper towels. Shiitakes have white spores and Chanterelles have yellow, sometimes pinkish, spores. Spore prints from these mushrooms will show up better on dark-color paper.

This takes a day, so choose your work area wisely.

SUPPLIES
- Mature gilled mushrooms
- Knife
- A glass or bowl larger than your mushroom caps
- Hairspray
- Paper or paper towels

1. Cut the stalk off as close to the gills as possible.

2. Place your cap gill-side down on the paper or paper towel. Use dark paper for mushrooms with light colored spores.

3. Cover with a bowl and leave undisturbed for 24 hours.

4. When it's time, carefully remove the bowl and cap.

5. Spray your print with hairspray to preserve the pattern.

FUNGI PERFECTI® MUSHROOM FUN TIME ACTIVITY BOOK

MAKE STEM BUTT SPAWN

Paul Stamets, founder of Fungi Perfecti, recommends using stem butts to grow mushrooms. Rhizomorphs (dangling string-like mycelial strands) radiating around the base of certain mushrooms generate new mycelium in the right conditions. You just need cardboard, a nice shady spot on the ground, and a little patience.

Eventually you'll have cardboard spawn, which you place—mycelium-covered side down—in piles of wood chips or newspaper in your garden to start a mushroom patch. Learn more about growing mushrooms in Paul's book *Mycelium Running: How Mushrooms Can Help Save The World* (2005).

Garden Giants (*Stropharia rugoso-annulata*) and Oyster (*Pleurotus*) mushrooms are great choices for stem butt cultivation.

Stropharia Rugoso-Annulata rhizomorphs

Choose a shady spot on the ground for this.

SUPPLIES
- Freshly gathered stem butts with rhizomorphs
- Cardboard soaked in water and peeled apart to expose corrugations
- A box, trunk, or container with a cover
- Wood chips

1. Put a piece of soaked cardboard in the container. Place stem butts roughly 16 inches apart on corrugations. Keep adding layers of cardboard and more stem butts, but leave room at the top!

2. Cover last layer of cardboard with wood chips. Mist with water. Cover your box or container.

3. Incubate for 4–8 months and mist regularly. Keep on the ground in a shady spot until the pieces of cardboard are ready to transfer to a growing area.

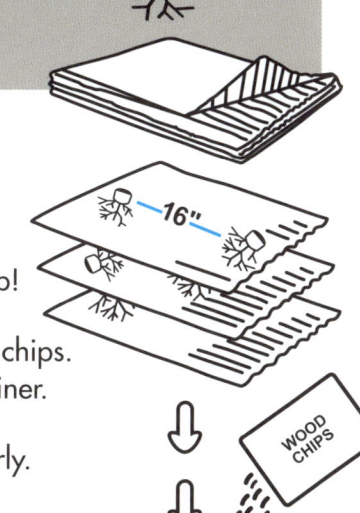

FUNGI PERFECTI® *Makers of Host Defense Mushrooms* | FUNGI.COM

DON'T FORAGE IN THE DARK!

Here in the Northwest, October rains beckon mush-roamers into the woods in search of delicious Matsutakes and Chanterelles. As night falls, foragers are often surprised by the ghostly glow of fungal beings in the dark. Once regarded as magic, bioluminescent fungi glow up the forests at night.

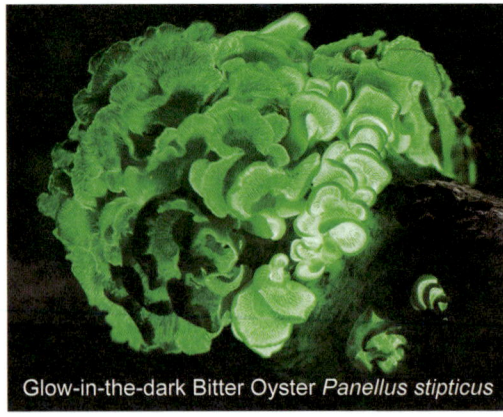

Glow-in-the-dark Bitter Oyster *Panellus stipticus*

There are over 30 different types of glow-in-the-dark mushrooms! The orange Jack-O'-Lantern mushroom (*Omphalotus olearius*) is poisonous and looks a lot like the delicious edible Chanterelle mushroom. This is why it's important to properly identify fungi. Before you munch down on a wild mushroom, take care to correctly identify it!

CHANTERELLE
One of the most popular wild mushrooms

JACK-O'-LANTERN
A very poisonous, glow-in-the-dark mushroom

Mushroom hunters rely on spore prints, guide books, phone apps, and experts to identify the species they find. If you're interested in foraging, find a mushroom group in your area through the North American Mycological Association. It's a great way to meet other mycophiles who will help you safely hunt for, and identify, wild mushrooms.

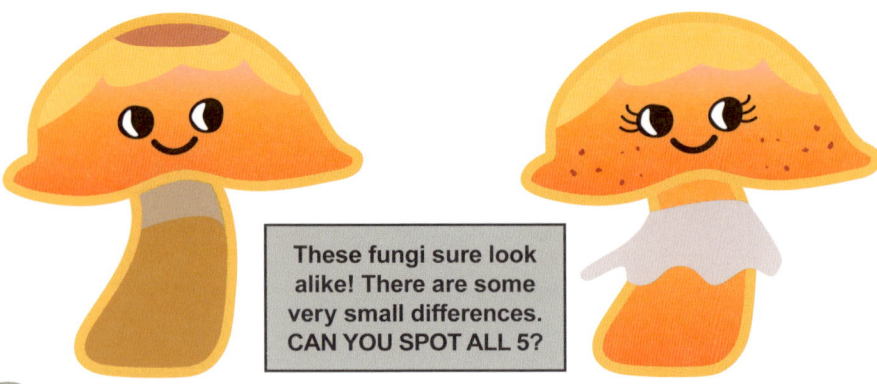

These fungi sure look alike! There are some very small differences. CAN YOU SPOT ALL 5?

FUNGI PERFECTI® MUSHROOM FUN TIME ACTIVITY BOOK

ANCIENT MUSH-TORY

Iceman Fired Up Mushrooms
Humans have used mushrooms for thousands of years. Oetzi the Iceman, a 5,300-year-old mummy of a Neolithic man discovered in the Alps, carried a string of dried Birch Polypores as well as other mushrooms possibly for medicine or tinder to start fires.

Dancing with Mushrooms
Among the ancient Tassili cave paintings in Algeria there are many images of dancers carrying mushrooms. *Tassili Mushroom Man Matalem-Amazar* depicts a shaman wearing a bee-shaped mask and dates back to 5000 B.C.

Humongous Glowing Fungus!
The largest known organism in the world by area is a Honey mushroom (*Armillaria ostoyae*) in Oregon's Blue Mountains. The colony is over 2,000 acres and around 2,400 years old. Some think it weighs 35,000 tons (about 3,900 T-Rex dinosaurs). Honey mushrooms are also one of the many varieties of fungi that glow in the dark! The phenomena has been called foxfire, will-o'-the-wisp, and faerie fire, and has been noted in the writings of Aristotle and Pliny the Elder.

Giant Prehistoric Mushroom Trees?
About 430 to 360 million years ago, Prototaxites, a giant tree-like fungi, was the largest land-dwelling organism. Fossils show us that they formed large trunk-like structures up to 3 feet wide and 26 feet tall!

(wikipedia.org/wiki/Fungus)

Retallack / CC BY-SA.

WHAT'S IN A NAME?

Have you ever wondered why scientists say *Canis familiaris*, when "Dog" is so much easier? Scientific names (binomial nomenclature) are part of a system that makes it so you can correctly identify things in nature—no matter what language you speak. Living things are categorized into **Kingdoms** (Plants, Animals, Fungi), then into a bunch of other subcategories, and finally a **Genus** (*Felis* for cats, *Canis* for dogs) and **Species** (*Canis familiaris* for domestic dogs or *Canis lupus* for wolves). In a binomial, the first part tells you what Genus it belongs to, and the second part identifies the Species.

Mushrooms with names like *Hypsizygus*, *Pleurotus*, or *Xeromphalina* sure can tie one's tongue, but don't fret—there isn't really one way to pronounce them (they're ancient Greek and Latin!). When in doubt, just follow Paul Stamets' advice: "Say it with authority!"

It also helps to pretend you're a wizard casting a spell: *It's not grifola frondosa. It's Gri-fola fron-doh-sa!*

Let's learn some *Latin* by breaking down the scientific names of some of our favorite mushrooms.

MAITAKE — *Grifola frondosa*
- grifola = braided fungus
- frond = a leaf
- osa = abundance

"Braided Fungus with a lot of Leaves"

TURKEY TAIL — *Trametes versicolor*
- tram = thin
- etes = one who is
- versicolor = multicolored

Trah-mee-teez ver-si-color!

"Thin One Who is Multicolored Mushroom"

REISHI — *Ganoderma lucidum*
- ganos = brightness
- derma = skin
- lucid = glossy or shiny

Gan-oh-der-muh loo-see-dum!

"Bright-Skinned Shiny Mushroom"

LION'S MANE — *Hericium erinaceus*
- hericium = hedgehog
- erinaceus = hedgehog

Hair-a-see-um air-inna-see-us!

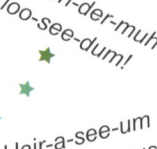

"Hedgehog Hedgehog Mushroom"

FUNGI PERFECTI® MUSHROOM FUN TIME ACTIVITY BOOK

> If I were a mushroom, I'd be a *Versicolorosa felinastra* "Abundantly Multicolored Cat Star" mushroom.

Draw your own mushroom and give it a scientific name!
Here's a few words to help you build a binomial. *(Latin / Greek)*

red	*cydonia*	**beautiful**	*formosa / panemorfi*	**bear**	*ursinus / arctos*
crimson	*phoenix*	**brave**	*animosus*	**bird**	*avis*
pink	*rosea*	**compact**	*compacta*	**book**	*liber*
orange	*aurantiacus*	**confusing**	*pseudo*	**cat-like**	*felina / gata*
yellow	*luteum / citrin*	**deceptive**	*fallax*	**crow**	*corvum / koráki*
gold	*chrys*	**delicate**	*subtilissima*	**dove**	*columba*
green	*vireo / khloros*	**love**	*phila / agapi*	**earth**	*geo*
blue	*cyan*	**scurfy**	*furfuracea*	**goat**	*capri / gida*
purple	*purpur / violéta*	**sensitive**	*sensibilis*	**horse**	*hippo / alogo*
silvery	*argenteus*	**silly**	*ridiculum*	**leaf**	*phyllon / fyllo*
bearded	*barbatus*	**smart**	*captiosus*	**mushroom**	*hydnum / manitari*
jelly-like	*gelatino*	**strange**	*novis / paraxenos*	**octopus**	*polypus*
flowing hair	*chaetes*	**water-loving**	*aquaticus*	**owl**	*ibis*
fluffy	*crinitus / afrato*	**wild**	*ferox / agrios*	**river**	*potamos*
milky	*lactarius*	**largest**	*maximus / megalo*	**snake**	*orphis*
spined	*chordatus*	**many forms**	*polymorpha*	**star**	*stella / astro*
spotted	*maculatus*	**heart-shaped**	*cordatus / kardia*	**squid**	*lolligo / kalamari*
velvety	*velut*	**short**	*brevipes / mikros*	**tiger**	*tigrinus / tigri*
woolly	*lachnos*	**smallest**	*minimus*	**tongue**	*glossum*
wrinkled	*caperatus*	**split**	*schizo*	**wood**	*xylon*

FUNGI PERFECTI *Makers of Host Defense Mushrooms* | **FUNGI.COM**

MUSHY NAME MATCH

There's probably a million *undiscovered* species of fungi. Of the 120,000 known species, each has a binomial and a common name sometimes inspired by its color, texture, shape, or smell. See if you can match the common names of these mushrooms with their Latin binomials.
Even if you know some Latin, it's not easy!

INDIGO MILKCAP	*Agaricus arvensis*
TOOTHED JELLY	*Sparassis crispa*
JELLY EAR	*Flammulina velutipes*
GIANT PUFFBALL	*Craterellus cornucopioides*
ENOKITAKE	*Lactarius indigo*
CAULIFLOWER MUSHROOM	*Calvatia gigantea*
HORSE MUSHROOM	*Marasmius oreades*
TRUMPET OF DEATH	*Pseudohydnum gelatinosum*
FAIRY RING MUSHROOM	*Auricularia auricula*

What is Shiitake's favorite drink?

FANTASTIC WORD SEARCH

We've hidden a lot of fungi below and still had mush room for words all you future mycologists should know. Can you find them all?

```
G A C F R E S S H E K U Y T O Z D B D M P X T
M K R E I S H I Q Y Z C O F Z K G I O Y D B E
S O F T O M Z K D L Y N U I O S M W Q C X E N
H F R S Y I W O Z Z K N Q L T R G M R O D K O
I L D M F S B R L L J O N E T I A B O L Y Y K
I J O Y S T E R A A L R M X T I K G Z O M H I
T G I F I V C H M B P A S C K N V W E G S E T
A V W U E O P G A Y T Q E X O V J A M Y R Z A
K Z R K Z Y Y P M S F F B C H A G A T O K W K
E F X P H Q H X E N R K P E P U K V P I L Y E
B B F X H O S T D E F E N S E I I S S C O O L
E E G J D S M S P J L I F W P T N H N W I N C
R M E O C M A I T A K E S I C U G A O S P W M
N Y T F R M G D P X X G B O V R S G C T O D U
C O V H R N P R I M O R I D A K T A O A L B S
L I Q F U I L I O N S M A N E E R R R Y L H
I O S F A T E P I N H E A D F Y O I D S P K R
D P V C M R R N W V F O R E S T P K Y H O J O
B F Z A Q D T A D P M E S I M A H O C I R X O
L J W P U W P J O L L I B V Q I A N E P E I M
M K C C G S J B N S Y A O S I L R S P M A N R
R O Y A L S U N B L A Z E I T V I B S L B W L
M I N O C U L A T E A E R K X I A M A D O U P
```

AGARIKON	~~FORAGE~~	MESIMA	SHIITAKE	STARSHIP
AMADOU	FOREST	MUSHROOM	SPAWN	TURKEY TAIL
BEE FRIENDLY	FRUITBODY	MYCOLOGY	SPORE	
CAP	FUNGI PERFECTI	OYSTER	STAMETS	
CHAGA	HOST DEFENSE	PINHEAD		
CONK	HYPHAL KNOT	POLYPORE		
CORDYCEPS	INOCULATE	PRIMORDIA		
CULTIVATION	LION'S MANE	REISHI		
ENOKITAKE	MAITAKE	ROYAL SUN BLAZEI		

A cappuccino!

POLYPORE PAPER

Mushrooms break things down and create soil for plants to grow, but mushrooms can make more than dirt! Some of the tougher polypore mushrooms (they have pores instead of gills) can be made into paper.*

Turkey Tail (*Trametes versicolor*) is one of the most easily identifiable mushrooms in the world. There are many varieties of this multi-colored, fan-shaped bracket fungi. The fruitbodies have alternating dark and pale zones with a white margin, and are notably tough and woody because their cell walls are made of chitin. Chitin is similar to cellulose from trees, which is used to make paper.

Choose a space where water is available and big messes are allowed.

SUPPLIES
- Turkey Tail mushrooms
- Towels or fabric
- Cooking oil spray
- Blender
- Large basin or tub
- Plexiglass
- Paper-making screens or molds, or glass bowl

1. Cut Turkey Tail mushrooms into small pieces. Freeze first to kill any insect larvae lurking within, then soak in water for a few days (refresh the water occasionally).

2. Put mushrooms in the blender with a little water. Pulse until you have a fluffy pulp. Transfer to a large container.

3. Shape a film of pulp onto a paper-making screen, or use a glass bowl as a mold (use a cooking oil spray to prevent sticking).

4. Flip paper onto old towels or blankets to absorb excess water. When it's mostly dry, use plexiglass to remove paper from padding. Leave it in a sunny place for a few days to finish drying out.

5. Freeze the paper for a few days before you use it.

* Check out Miriam C. Rice's book *Mushrooms for Dyes, Paper, Pigments and Myco-Stix* for detailed instructions.

MUSHROOM INK

Did you know that you can make your own ink with mushrooms? Hundreds of years ago *Coprinus* mushrooms were used to make ink for writing letters and drawing. *Coprinus* mushrooms, sometimes called Inky Caps or Shaggy Manes, are pretty easy to find in the spring or fall popping up on lawns all over Europe and North America.

At first, they look like fuzzy beige drumsticks. *Coprinus* mushrooms quickly decompose into an inky black goo—melting like the Wicked Witch of the West. This opens up their tightly packed gills so spores can be released.

SUPPLIES

- Mature Coprinus mushrooms starting to goo up
- 2 Jars with lids (one large and one small)
- Pot and permission/help to use the stove
- Strainer
- Essential oil *(optional)*

1. Put Inky Caps into a jar for a couple of days. They will decompose into a stinky, gloppy goo.

2. Strain the goo to remove any chunky parts and dirt. Put the liquid into a pot.

3. Reduce the liquid by boiling it for a few minutes. Transfer the ink to a small jar when it has cooled.

4. Make your ink smell better by adding a few drops of essential oil (try rosemary or lavender).

5. Your mushroom ink is best used with a fine paintbrush or quill.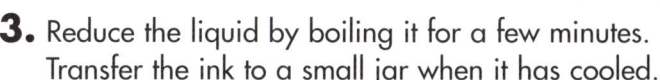

AMAZING INTELLIGENCE

Fungi have been called "Earth's natural internet." Plants and trees plugged into a mycelial network have faster and stronger immune systems and recieve extra nutrients and protection. Seems like a pretty smart system.

We used to think slime molds were fungi, as they appear like them in one stage of their life. We have so much more to learn! To all the organisms formerly known as fungi, we salute you! A slimy yellow former-fungus, *Physarum polycephalum*, found the shortest distance to a yummy oat flake in a maze just like this one. Scientists think it's evidence of cellular intelligence.

PERFECTI CROSSWORD

ACROSS

1. The study of fungi
5. An Olympia, Washington company dedicated to helping bees, trees, people, and planet
8. The largest known organism in the world by area (also what bees make)
9. A baby mushroom
10. The underside area of a mushroom cap, and fish have them too!
11. Underground branches that absorb nutrients for a mushroom

DOWN

2. A popular edible mushroom, and also a salt-water bivalve mollusk
3. Ancient, gigantic tree-like fungi
4. A tiny dust-like particle released from gills that grows new mushrooms
6. To go hunting for mushrooms
7. The dome-shaped top of a mushroom
12. A dark liquid you can make with *Coprinus* mushrooms

ANSWERS

HIDDEN CREATURES PAGES 2 & 3

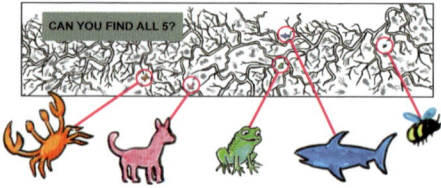

WORD SEARCH PAGE 15

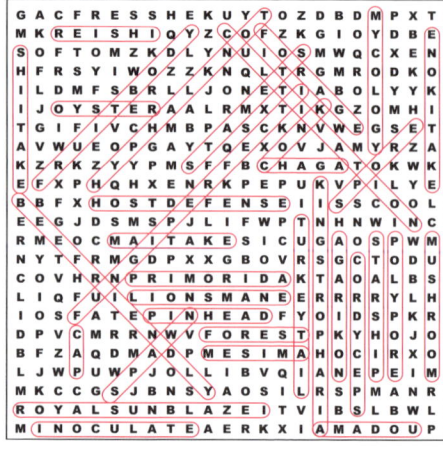

SPOT THE DIFFERENCES PAGE 8

1. Red spot on top of cap
2. Eyelashes
3. Small spots on bottom of cap
4. Color of stem
5. Annulus / Ring

MYCELIUM MAZE PAGE 9

MUSHY NAME MATCH PAGE 14

INDIGO MILKCAP *Lactarius indigo*
TOOTHED JELLY *Pseudohydnum gelatinosum*
JELLY EAR *Auricularia auricula*
GIANT PUFFBALL *Calvatia gigantea*
ENOKITAKE *Flammulina velutipes*
CAULIFLOWER MUSHROOM *Sparassis crispa*
HORSE MUSHROOM *Agaricus arvensis*
TRUMPET OF DEATH *Craterellus cornucopioides*
FAIRY RING MUSHROOM *Marasmius oreades*

CROSSWORD PAGE 19

ACROSS: 1. Mycology, 5. Fungi Perfecti, 8. Honey; 9. Pinhead, 10. Gill, 11. Mycelium.
DOWN: 2. Oyster, 3. Prototaxites, 4. Spore, 6. Forage, 7. Cap, 12. Ink.

HIDDEN TIGER & AMAZING INTELLIGENCE PAGE 18

© 2020 Paul E. Stamets & Fungi Perfecti, LLC. All rights reserved.
Many of Fungi Perfecti's myconauts contributed to the creation of this activity book!

BOOKS & REFERENCES

Growing Gourmet and Medicinal Mushrooms (1993) and *Mycelium Running: How Mushrooms Can Help Save The World* (2005) by Paul Stamets

Mushrooms for Dyes, Paper, Pigments & Myco-Stix by Miriam C. Rice

Fungal Luminescence and *What's In A Name?* by Tristain Woodsmith at fungi.com

wikipedia.org/wiki/Fungus

ILLUSTRATIONS

"What is Mushroom Mycelium?" YouTube, uploaded by After Skool / Mark Woods, March 10, 2020. Additional screenshots from After Skool / Mark Woods.

Cordypillar & Chibiitake by Sophie Hendry

Other mushrooms and illustrations by Alyssa Parker

FSC RA-COC-001386
Printed on recycled content paper

FUNGI PERFECTI® MUSHROOM FUN TIME ACTIVITY BOOK